Principia-Logico

Written by R E. Ford
Photography by Lela Maisy

Other work by R E. Ford

Justified Dreaminess
The Humility of Three Carvers
Dancing with Strangers
20 Songs of Innocence and a Song of Vitality
Self-Reconcile
Nova Ordem Mundial
Welcoming Freedom
Amet Verum
Jailarbeiten: Phaenomenoligie auf The Body, Mind,
Soul, Spirit and Hallow Mornings Coming Fast
When Knowing, Imagination, and Humility
Meet—Good Wins: The Poems for Osiris
Spirit and Soul
Nihilism and Time
Capulets
Naked Consciousness (forthcoming winter 2022)

Principia-Logico

Nymphs Lit
146 Thornleigh CT
Brownsburg, IN

IG: @thephantombrother
email: reford19@gmail.com

For Delp, he cool

Preface

In Quantum Theorum of reformation of existential mathematics, it infers at this dawn of its predication–we can no longer be breached as singularity in our affirmations to be redefined in a general directive of a young artist there are post-pandemic realities of what is magical and beneficial in survival due to odds of a generation that was alas educated on what biology and physiology are.

The pivotation of high end legacy in biophysics will rise in the ensuing century as the political side of such in ethics is at the turning point of facing trial in ethical fallacies of what can and should be news in the world of *pathos, logos, and ethos* in the ideas of the human body and genome.

The complete context of arguable demeanor is that I have answers I have given without credit to denounce plagiarism and reform it as inspiration and great influence of theory–not human practice. The future auguries of human compassion will see how the physiology studies in biology are indeed the faithful's world of our body such as our souls and spirits in human capacity. This merit of deep identity alas has surmounted to where new course words can be relative to infiltration of a rise in character as the studious pick metacognitive in debate over what is typically a dividing line of divine and the indoctrination against even secular belief.

We preface lack of identity in course contact of how one is spotted in their own becoming of whom they can be—independent of all outside influence in the Rilkean idea in neuroscience metacognition reflection of triggering past dimension as inward understanding of psychological freedom and how basic quantum theories are triggered at humanistic molecular chemistry or biology. One used in substance control intake—the other in actual neuroscience of molecular psychology. In that moment such is studied on solipsisms of whom the selfish and greed of a capitalist society—the individual will acclaim proprietorship of the inward reflecting out as we analyze the quantitative data of quantum psychology from in depth discovery of—metacognitive physiology, cognitive, molecular at whole, and cardiology of remaining away from the sleeve in this treatise.

R E. Ford
October 14th 2021
Brownsburg, IN

Table of Contents

"Wir reden über Poesie so abstract, weil wir alle schlechte Dichter zu sein pflegen."—Friederich Nietzche

Part One:
Amet Verum

1.0
And then, there I was, alone again. It was not from which I always was in search of absolution as before. It was that I could describe how I had attained it.

1.01
These examinations were not always pleasant before, inside of it, or after. The startled beginnings of madness.

1.011
Then innocence from repression. There were
drawn out ways to redefine each new memory
in the perception of how questioning the
necessities that were no longer seen as such.

1.012
It was aloneness in thought. Not
loneliness—the derivative of despair from being
lost in the search of absolution.

1.013
There is a secret that it was foreshadowing.

1.014
The very nurturing wisdom that unwound
inside, before there was even a need to search
for absolution, but with that, was not
addressing others.

1.015
That is where becoming stuck in delusion arose
from profound complexity in childhood's end
in insanity passed on inherently from
generations that did not ask these questions.

1.1
A self-evaluating claim in reassessing what
values there were—a climb upward was slowly
taken back by the turn inward.

1.11
Absurdists saw this as down is why many do
not know how the journey to psychological
freedom may even begin to happen.

1.2
Where you are psychologically free, there is
creation from journeying in to be removed
from the physical world with its burden of pain
and to understand the mind with its flaws and
pain as well.

1.21
With the incorporation of both begins the
process to go from perceptions that are outside
the normal *one* that most have.

1.22
Slowly, there is nothing inherent except what is
spiritual and biological.

1.23
The physical world's outside demands are no
longer intriguing. This is the split between
loneliness and aloneness.

1.3
There I was at consciousness.

1.301
Those first bright moments of nonliving or thought without question—invasion of the physical world.

1.31
Clear objectification is the incoherence of separate minds pushing at different end goals through what is their perception in what is possible. To not rebel with an opposing view, but to claim it your own in injustice is to not move forward.

1.311
This is the surface; you have a view, but it becomes irrelevant. Progress and digress move at slower speeds, but the tilt in one over the other is magnified much greater, because it is abnormal to one who does not question opposing views.

1.32
The worst is to have no view.

1.321
The conclusion is that to take your own view is radical.

1.322
It is not entirely easy to do if you don't
transcend or descend from the surface in some
manner.

1.33
Just as depth is under, surface is steady, the
movement up without either is blinded by the
light.

1.34
There at the center of everything within the
history of intelligence is what the questioner
and listener can learn, what the questioner can
learn, and what the listener can learn.

1.341
The equivalency in the historic sense of
psychology—superego, ego, and id—where
reevaluating has been done. To escape either in
pure psychological freedom you must be able to
traverse all three at will. In order to do so, the
journey within eventually must be readjusted
in order for there to be no social pedagogy.

1.342
Psychological freedom arises from finding
depths deeper than others, if you can find
abstraction's correlation to the surface; if not is
where genius turns to insanity.

1.35
The quintessence of childhood is to take it as
amet verum.

1.36
Where the truism of childhood disengages from
logic is the only way to put this ideal at center.

1.361
Few process the past or even dictate the
present with clarity of how it handled pleasure
in its innocence, which is how such a term
developed.

1.362
The ploy of the world at whole is an
unresolvable force to proceed further in a way
to only declare ramifications in physical
possessions.

1.363
This view is the digression of true value. All
points to and from *amet verum* are not always
personal consciousness, but consciousness
unfolding in the ways that strip individuality
from where personal spirituality and
self-knowledge can be solely—up to the
individual.

1.37

To harness such is in *amet verum*. In where its origins are in childhood, but slowly make its way back in only moments in most lives.

1.38

The individuality of pain is not given proper justice as that—perceiving it as individuality. The consciousness of being an individual is not justified by physical liberty. Rather a misconception that has been forever been part of the world's injustice to *amet verum*.

1.39

There is no journey within, in which every point it will force you further from *amet verum,* there is no way to see it in its true form outside of childhood—to even be aware that it was developed in innocence at birth.

1.391

The essence of psychological freedom is preceded by the known existence of *amet verum*.

1.4

Then what is *amet verum*—in the simplest of ways, untrained knowledge, fun, and truth.

1.41

From that is how each individual can asses the outlines themselves.

1.411

With any word there is the technical definition and the creative.

1.412

The difference lies in one is for scholars, and one is for free-thinkers.

1.413

Free-thinking without some scholarly work makes the individual not able to be psychologically free.

1.42

Mental capacity grows from scholarly work.

1.421

Free-thinking is not always the mind realizing depth, surface, and above.

1.43

Scholarly work struggles the most with anything outside the surface. They are not opposites. They are the ways for an individual to know their *amet verum*.

1.44
There was a precise reason the term found itself within literature as all do.

1.45
Evolution with emotion with everything is becoming anew.

1,451
There are more experiencing different emotions from the complexity of how the mind is accumulating more information from all the corridors of its possible expansion.

1.452
This is where the psychologically aware take technology and dispose of it's inadequate nature within the masses, and mend it as the way to guide them to their personal truths.

1.453
It is not easy.

1.454
Being psychologically aware can only allow the individual to take on the complicated task of their push towards these emotive realizations taking place.

1.46
There are, throughout history, many takes on as I call it—supernatural phenomena.

1.47
In many accounts the original nature of it gets replaced or renewed as time goes forward.

1.5
There should be no detracting from any religious outlooks or universal truth one finds within it, but the common way to go about it is to hold superior standards in some.

1.51
Where you remove religion and leave it up to the individual, there is always a degree of misinterpretation—question without answer.

1.511
Where social pedagogy is not the way it can truly harness the power of the hermeneutics within the literature of Holy Scripts, there to a degree must be a renewed moved towards refining the individual nature we have discovered—to know that where the purists see the past that this turn towards individuality is not as scriptures are rejoicing supernatural phenomena is a much greater flaw they in themselves must address as their own individuality on the stance.

1.512
It is on both sides that the individual is merely happening.

1.513
Both do it more blindly than they want to admit. The purists are the only ones not moving somewhere within nihilism's domain—the move to and from finding what values you really have.

1.52
The blind nihilism in that, is just as passé as the purists view on individual views of supernatural phenomena.

1.521
A much more self-aware way of looking at it is the questioning of what an individual must go through in removing *amet verum* in order to dive into the depths of perception.

1.53
There are things known and unknown to many.

1.54
The constraints on most individuals is that they do not go in search of ideas that contradict their stances on their beliefs.

1.541
Most only read to reiterate how they already perceive.

1.542
There is the biggest flaw within education on any level—self or formal—that you do not know what is exactly making history in your moment in time, outside of the major events of that are altogether impossible not to; where so many only see the present by what is handed to them in that, as opposed to trends that are too far ahead of the curve to have proper attention directed towards.

1.543
For this is where it is the truest scholar's biggest obligation to society—to look beyond what is presented as there, instead cultivate the trends of the past as analyzing their movements in history, to know what relevant ideas could possibly develop with how they have in history.

1.544
The vastness of that problem now is that every bit of information, technology, population, and ideas are expanding much more rapidly than ever before. The true scholar would inherit from knowing history that that is how it has always been.

1.545
History has in and of itself always been a
distraction from well-known contemporaries;
who at any point in time, slowly fade in
history's shadow from not knowing and
projecting them in the moment irrelevant
compared to the very trends that will define the
present and move more in the future.

1.546
This is where there are very few in history who
attained psychological freedom, because as you
move away from *amet verum* in life it must be
in rebellion with moving against society to
properly place the interest of society at center
from tearing apart their fallacies in that
moment in history on how it detracts from
what is the greater good for everyone in order
to justify a better future free from the struggle
of that time—to project a brighter future from
the obscurity of seeing everything as is, on an
infinite view of history—not linear.

1.6
There are ways in which unknowns become
readily available to individuals. Some are
through visions and long, immense connection
to supernatural phenomena.

1.61
That correlation to the surface or
understanding on that level is much harder to
bring forward.

1.611
The idea of attaining the unknown through
poetic reasoning is generally not understood in
the realms that do not allow it to ever be
brought down from its cloud or up from its
depth.

1.612
That is where, eventually, transition from a free
thinking scholar to a true scholar can allow the
rest of society to follow the thought of what
poets have forever been trying to express
without the actual justice that coincides with
poetry.

1.613
There must, ultimately, be a foundation of
further examination beyond creative work with
invariable merit to how the individual looks at
it their way.

1.614
It requires the fundamentals to never declare
one is worse or one is better, but to put the
entire vision from their depth or cloud in work
beyond poetry.

1.615
There are many fierce poets throughout history
that have climbed these peaks or plunged these
depths, but to bring it back to the surface in a
more universal way is the task of a true scholar.

1.616
Their reasoning must be not in the name of
truth, but to allow life to proceed on after
poetic examination.

1.617
This task of the true scholar is done in life away
from amet verum in order for more to
experience it on a surface level without them
being unawakened by the triad within how life
works, and not allowing them to live on
autopilot as so many do, even in their moments
of amet verum.

1.62
The continued expression that is heard is that
everyone's got something—a story or past of
suffering.

1.63
There are few who take the depth of perception
from it and transition it to the view that every
new moment away from it is redemption.

1.631

In those times of suffering where no end is in sight, not everyone gets to redeem themselves then to take on everything as new in a childlike wonder of how enriching life is on the other-side of moving on from it.

1.632

This is something that scholars cannot teach or anyone who has not experienced the devastation of life without meaning in facing suffering without voice.

1.64

So then the line is drawn—a true scholar can only be if there is thorough understanding of suffering and understanding of its place in history. In these moments away from *amet verum* with being a scholar it is essential for expression to take place with any medium found sufficient.

1.641

Amet verum is not easy in the darkest moments of emotive existence.

1.642

Amet verum is the daring task that can only be done if it is put at the center of your values; where everything else derives from that value.

1.643
This has forever been the biggest fallacy in the modern era—to take art and philosophy away from the center of values.

1.65
A true scholar must learn art at all levels to know what art is in his era. It is up to the individual to start a change within—before external.

1.651
Expressing oneself doesn't have to be anything other than to themselves—the others around them will eventually inherit this change from them, if there is no one around, it is possible to ask—why?

1.652
No change is small change as anything that permits oneself to see beyond what they have always known is in and of itself significant, and a step not many ever take in the first place.

1.653
These are essentials towards psychological freedom—more inherent than perceived.

1.654
The knowledge and expression of individuals is seen as a threat to power.

1.655
A true scholar doesn't follow suit with either expression or power, rather must reach for some universal truth of his era with both.

1.66
In language, art, and power is only through the voice of the people, if it is to be properly put and just. These are not things that come quick, or at all for most.

1.661
An entire life can be spent in peculiar undertakings to explain such movements in life that endures to old age—truth may never be found in it.

1.67
In the study of language comes emotion and the mind's way in individuality of how to use it outside of art or power; that is how psychology is formed.

1.7
Psychology is the transition in the human mind to connect language to emotion that properly expresses how that relates to day to day environmental events.

1.71
Through this is either lack of or more than.
This is the middle point of psychology. There
are broadened ways of putting it in the science
or much less technical.

1.72
This is because society has become
complicated.

1.721
Society is complicated, because many do not
stop and think.

1.722
Then when society has to, their mind
overworks, and even insane ideas turn into
something they address in these situations.

1.73
Then vice-versa, the people that are in tune
with their mind and psychology intuitively
realize this in other's perspectives.

1.74
The turn from existentialism in society was
developed through psychology.

1.741
With the vastness of science and study in the
field of it with all the derangement of the

mind—it replaced the crisis with therapy and medicine; it culturally removed it in a way to silence the extremely creative by dejecting their mind and emotions from the anger they felt by being objectified by monetary value as to what they inherently had to offer.

1.742
Slowly free thinking was a diagnosis for the ones that thought with their own mind in their own way.

1.75
Historically, every great thinker would've been on antipsychotics or antidepressants or both from the very idea that they were bipolar—is how it is perceived.

1.751
We silenced these people for generations.

1.752
These were the beginnings of psychological development.

1.76
The decline in the value in art and philosophy was replaced by the science of psychology. These are the very reasons the realizations of *amet verum* were made.

1.77
There must be a call for a new way to go about underdeveloped psychology.

Part Two:
The Evolution of Love

2.0
The flaw in therapy is that they want to tie it down to a science and find solutions like a math equation.

2.01
Therapists will have you show your work in stream of conscious speech trying to put the proper symbol in from a book they studied where each variable of emotion equates a solution—a measure of science, pharmaceuticals.

2.011
The human mind is not only much more complicated than any such algorithm that they teach you in math.

2.012
The mind is entirely different for each individual; where it takes a much more infinite understanding of logic than science or math itself.

2.013
Many also don't think about their emotions.

2.014
Emotions are felt, not processed through thought.

2.015
Many do not have the perception that no emotion is positive or negative—that anything felt is important to comprehend what it developed from and unravels ways to know yourself better.

2.016
It does take psychological awareness to process emotion this way, but a step is to understand emotion as learning—to perceive everything as an experience of emotion to eventually be processed in thought to learn about yourself.

2.017
The subjectivity of emotion in positive and negative ways they've developed is even in and of itself different in each individual as well—making any such projection about it is yet another way culture has removed individuality.

2.02
Culture has a vague way of love.

2.1

Love can be reckless.

2.11

Many sway within its pull from a plethora of ways it conquers.

2.12

The very haunting *esse* is found too often. It comes and flees.

2.13

The personality of the individuals experiencing the essential nature of love find the miracle of it.

2.14

The fleeting beginning is sometimes far too unknown in an endless question of why or such immodernity.

2.2

Psychological love isn't the same as psychological freedom.

2.21

Their split is defined by each individual. To restrain from lack of finding these values is where too many know the past is finished with this charade that was hardly touched on.

2.212
Some know, most don't. Sartre clearly pointed out that authentic love doesn't last. In that, you must replace authentic love with psychological love.

2.213
In order to do that there must be a deep self-awareness on both sides to fully embrace that the beginning is authentic, and to continue it—the psychological side of both must develop as one.

2.3
The split between psychological love and psychological freedom comes in the reality of realizing the latter is only attained through an immense search within and externally in order to understand both to full-extent.

2.31
In search of psychological freedom one must understand the self and others for flaws that need to be repaired.

2.311
Psychological love is found between lovers and both partners can explain how they gave up their freedom for the love they saw in the other.

2.312
Then the slow build of psychological love
begins.

2.313
Without that build resentment will come, if
love isn't fully developed. Many do this blindly,
but now it is needing to be explained so that
there are steps for both.

2.32
There is always free-will, but if you are not
trying towards psychological freedom or love
then danger can creep into the dark matter of
the mind or psyche. It is necessary to know love
or yourself fully.

2.321
There is an important message sent in
examining the self that allows a
boundary-driven path chiseled to
deconstructing bad habits.

2.322
There is a darkside and pure light inspiring
everyone.

2.323
When using *amet verum* to proper palace—the
key of love and freedom's tangling.

2.324
That is just it though—*amet verum*–the
defining *esse* so many never fully understood
the concept.

2.3241
From ancient to distant future it is important
to know now.

2.325
The gravitation, the beauty–where endless
reassertion of child-like wonder follows a
youthfulness too many have never embraced.

2.326
This examination of the self resurges with *amet
verum*. All the angles of psychological love and
psychological freedom bring comfort only in
true autonomy of *amet verum* which is where
one's curiosity builds attaching the two.

2.3261
Bountiful is the occasion when some bleed and
pass away without *vigilância*.

2.327
Then *amet verum* has been identified as the
answer between love and freedom.

2.3271

The more up close problem, how does an individual get to *amet verum* from any walk of life? The mundanity or mediocre pleasure or lifestyle of many is the problem *amet verum* can solve as well. People have a drastic-displeasure of getting stuck.

2.33

The people that get stuck find endless boredom as they grow older to become fluent in freedom is a challenge for too many of these people who only see life in a very simplistic way.

2.331

Burying pain always negates these insolent meanderings of mediocrity.

2.332

Mediocrity is the opposite of *amet verum*. In order to escape it—values or morality are at stake. Anything less than mediocrity is not an individual's fault.

2.34

It is the world's lack of reaching out and understanding psychological love.

2.35

Love's injustice is also haunting or pleasure.

2.351
The long list of its passing through millennials
has yet to be explained in its revisions of
evolution. That is where the evolution of love
slowly became reckless.

2.352
The intervention to manifest psychological love
is the beginning to explain its evolution.

2.353
So as time refrains strangely with love losing
too much. We now turn to the solution of
sustainable love, and its return in the psyche of
evolved longings.

2.36
Love had no depiction at the beginning.

2.361
If God was necessary to invent, so too was love.

2.37
To comprehend according to Adam and Eve in
its allegorical manner is to give place to God
and how chaos was first.

2.371
Freewill was granted.

2.372
Freewill is indifferent to love.

2.4
Then it must be asked how it is possible to see this evolution of love.

2.41
The standard of love has long been separated far too often from an unrealistic amount of force.

2.411
The most well-presented force is supernatural phenomena. The deity or deities that move symbolism in outrage to change.

2.42
Of course Hegel claimed, "God is dead."

2.421
That was an evolved moment, and that the highest power of existence had finally lost to humanity.

2.4211
Then as centuries later we're finally seeing what Darwinism developed. It was in the 19th century that the very term developed—science.

2.422
From there the evolution of love was drastic.
Somewhere along the evolution of genetics
affected the mind and psyche as well. Many
knew then that science defeated God.

2.43
Eventually the centuries of scandal flip then all
becomes undone.

2.431
The very irate motivation without psychological
freedom became invisible time after time.
Sitting high or low—noted prior—is the hardest
task for man to see, yet women's gentle *esse*
was, mostly, standing on the outside looking in
child bearing and loving the brutal nature of
men.

2.432
"Soul of a woman was created below," is how
Led Zeppelin put it.

2.433
This false claim was merely made from people
slowly seeing psychology was going to disprove
it. To take evolution and the nature of music
from the 1940s to the 1970s it was very similar
to how Greek theater developed. To prevent the
fall, there is a huge difference.

2.44
All genders have found women thoroughly
studied now.

2.441
They teach us psychological freedom and
psychological love.

2.45
The truth has been there for a while, but it is so
that it was too far ahead, or that we have failed
to embrace existentialism in math.

2.451
In which so many unclaimed ideas roam from
history that it would be much more beneficial if
this wasn't such a higher level
concept—existentialism or philosophy in
general. To think is an art form pried open
slowly by depth and above; it has been lost far
too long.

2.46
My case for *amet verum* is one in which it
greets an answer to the psychological that
answered existentialism.

2.47
Amet verum dodges the bullet of tracing steps
to brighter futures, and ascends mediocrity
through a clear lens. In that moment a star is

birthed as past and future fail to be replicated
in any way.

2.5
Absurdism is contained in knowing *amet
verum* is here, and as everything, it is evolving.

2.51
The absurdist is the contemporary mind.

2.52
It does not ponder the heavens. It knows the
heavens.

2.53
It accepts futurality of non-poetic nature.

2.54
The relativity of objectivity in physical
nature—yet aware of subjectivity of microscopic
physics in accordance with the universal,
individual is precinct of such a journey of
uniqueness in post-nihilistic cosmogony.

2.55
It is acceptance of chaos in humility of a
know-it-all ration.

2.56
It is he absurdist as an intellectual, rebellious
comedian is what uniform thought could derail

itself with the fear of uniform thought of such opposition.

2.57
Uniform thought does not permit the first coefficient in becoming aware that in relativity of subject matter—there is no uniform thought with each experience unique to madness.

2.6
Objective relativity in uniform thought that is that of post-modern.

2.61
To group in times of resolve is for the absurdist to see the non-absurd sneak off for anti-resolve.

2.62
The nihilist—not knowing the heavens—knowing dissolve culminates wealth.

2.63
Absurdists see the heavens entropy; so now they see the psychology of the nihilist as post-nihilist absurdist.

2.64
The absurdist can laugh at the absurd, but nihilism is pre-absurd, so absurdists must know it is indifferent to our inherent nature of

not being uniform with the nihilist—unaware of their objectivity in absurdists' subjective views.

2.65
A nihilist is militant. An absurdist is free-thinking and a contrarian to militant.

2.7
Absurdists believe militantism is from a lack of education.

Part Three:
Language of Science

3.1
one one>two
in compliance to
metaphysical is one two
then three is answer
in God complex
one four can be one

3.2
if one four are one
then four is never one two
zero is four
in regard to God complex
zero one two is two
not one one
one three>four
one three=two

3.21
in compliance to metaphysics
of being one in permeance
of dark light shift
one one three>three

3.22
three congruent to 1
yet non-compliant to 3

3.3
one and three semilar

3.31
God complex notes one
in addition it complies
in spectrum of three
one and one can be
three and three
or to take five as
numéric primitivism
are replicative in count

3.311
five is three or one
but not 5

3.312
numéric primitivism
divide is metric
of American count
integers are never irrational

3.32
God Complex is past tense

3.33
numéric primitivism is present tense

3.34
dark light shift is future

3.35
numetrics is all time

3.4
God complex is plurality
if Cartesian is one
God complex is one two
one two is metaphysical
of beginning for permeance

3.41
in historic count
one could not start
in one two there would be prior
of zero for other complexities

3.42
being-beyond-itself
is not always God

3.43
the complexity of phenomena
is sum is supernatural
in historic regard to count

3.44
Gods is a beginning
in historic count
to deter from plural sex
is different from complex

3.441
such phenomena is pretense
plural sex is ideal

3.442
historic count was ideal

3.443
if historic count was ideal
then historic report was false

3.444
false is not different than truth
it is an absolution

3.445
an absolution is God Complex
one two is a theory
one is three
in Catholic absolute

3.446
historic count is Catholic absolute

3.5
Catholic absolute has a God complex

3.51
in absolutes we defer truth

3.52
science is different than absolute

3.521
science retains the actuality of history
in actuality there is no hyperbole
in hyperbole are non-atheism
Catholic Mythology

3.522
Catholic Mythology is historic count

3.523
in what is the subject of whom
is what the integers in
other subjects are different
in these different natures
of historic count are Roman Numerals

3.524
Roman Numerals have no integer for zero or 0
or negative integers

3.525
God Complex is Roman Numerals

3.526
there is no negative in phenomena
of God Complex

3.527
four is null
ex why and ex ex
is start

(xx and xy in Roman Numerals other such way
is phonetics)

3.53
four parts create one two

3.54
one two is the equality
of four *null*

3.541
i ii is not one two
it is *unus duo*
iii is *tribus*
iv is quattor
(etc. in Latin)

3.542
Roman Numerals are nuversions
in numetrics

3.543
nuversions are starting integers
in numetrics as historic count

3.544
nuversions are inversions of infinity

3.55
An inversion uses calculus
to show opposing integers
of time permeance
and time shifts in metaphysics

3.551
a time shift shows
the differing factors
in time change in historic count

3.552
inversions apply to nuversions
and Arabic Numerals
(1,2,3,4...)

3.6
inversions apply to
English integers
(one, two, three, etc.)

3.61
inversions do not apply
to other languages
English is the root
with Latin for Scientific Language

3.62
Scientific Language is mathematical
as all languages are
the differential in it is complexity
for terms for varying fields.

3.621
The mathematics in Scientific Language
incorporate The Big Bang Theory
at atomic and all-encompassing level

3.622
Scientific Language is descriptive

3.623
the change in language
is from English to Scientific

3.7
Scientific Language has the
same alphabet as English
but it incorporates
both Latin for animals,
disease, and other former English names

3.71
the change from English to Science
is from the industrial and technological
revolutions in which
Computer Scientist changed grammar

3.711
Computer Scientist and Medical
have changed language to Scientific

3.712
if the art of English Playwrights
finalized such language with
rules to abide in the English Language
Socialized Media and Medicine
has its own indifference
to the rules of English

3.713
Computer Science is a part
of Scientific Language

3.72
numéric primitivism is in current
with Computer Science.

3.721
numéric primitivism is the algorithms
of search databases for research

3.722
bias in databases is media bias

3.723
media is scientifically literate

3.73
scientifically literate is dark light shift
that is a nuversion

3.74
when there is a nuversion
it must be finalized
in an inversion to know it goes on

3.741
a language as proliferate as Science
will endure transfinitely

3.742
to finalize Scientific Language
there must be universal comprehension
Homo Sapiens is humans etc.

Latin will be mixed in
but there are no capitals
or punctuation in Science

3.743
without punctuation in Language
was the directive of Computers

3.75
more humans than ever wrote
improper to English
in that a new language was formed

3.76
new language is formed
when life advances

Part Four:
Post-Nihilism

4.1

Contemporary art is ahead of post-modern science.

4.11

The answers of the existentialist have been discovered as anxiety being the artist's will.

4.111

In post-modern science, the doctor has remedies of theory, but not the scientific breakdown readily available for the contemporary movement available to fully understand.

4.112

These quantum theories are the basis for a breakdown to lift the science past merely nurses or medical professionals. To say a word of science is foreign to many—to educate a population of only few being entirely illiterate into scientifically literate is to end the war against the brilliantly inclined appearingly mentally ill—whom are merely the ones healthcare professionals fail at seeing as the artist of not just modernity, post-modernity, but the ones in contemporary times whom are self-aware of bad connotations of delinquency in which American Sciences fail at giving creative and doctoral advice.

4.12
Quantum theory is the final bough of saying the doctors gave postmodern mind bad advice as Alan Sokol has proved, and us in contemporary standards are now ahead of the curve–to say what we do not negate mental illness–we know it was an attack on the brilliant minds in American culture.

4.13
Quantitatively–Cantor, Dedekind, Russell, and Wallace were merely nihilisms worst nightmare.

4.14
The intimidation of authoritative identity of post-modern theoretical Quantum Sciences, the molecular level is under the scrutiny of academic dishonesty, and in mainstream psychology it is in generally deeply inherent that self-help books do not do any of explaining how Quantum theory at molecular psychology level can deviate the separate identities of mind and the brain.

4.15
The mind is metaphysical and metacognitive, the formation of phenomenology.

4.151
The functioning of the brain with melatonin
and serotonin uptake is on physical chemical
makeup–in which the theoretics of psychology
can be proven on molecular chemistry and
physiology makeup.

4.16
It is important to deviate psychology and
biology with one mentality of the brain and the
mind as opposed to the outside influence of
nurture and nature in Biology.

4.17
Biology is independent of physiology of the
human body and genome.

4.171
The double helix strands of DNA are
physiology. Biology is your another human,
dog, cat etc.

4.172
There is sociology which biosphere is more
remnant of the idea to divide psychological,
physiology(anatomy as a synonym) and the
sociology being nurture–nature is biology.

4.2
Quantum physics is on a molecular level, to see
it as biophysics is where many do understand
Astrophysics as astrology and physics. To be
fully in tune, one must know quantum makeup.
To dive into quantum physics is physiology
with quantum chemistry in chemical collusion
of what is the body to the combination of
quantum effect on an individual moment in
restoration of human capacity for
understanding a quantum leap is rather not
astrophysical.

4.21
Quantum physics is miniscule compared to the
transfinite capacity of the universe.

4.22
These differences are where we can watch the
quarks interact at molecular level in
physics—which is quantum.

4.23
Watching a comet be passerby or the identity of
a shooting star are astrophysics. Quantum
inferential leaping is fictitious. To understand a
human quantum is merely study the cell and its
intricacies on molecular biology breakdown of
such small quantum of mRNA with protein
synthesis in the ribosomes of the endoplasmic
reticulum—such is a bore for biophysics which

correlate to what technical quantum physics is in human form.

4.231
Past lives and transmutation are for the spiritual and artistic journey—it is not quantum or biophysics.

4.24
These identities of past lives are visceral to the dreamers of vivid *deja vu*.

4.241
These humans in the now, do not need to infer others of such lackluster allowance. To believe in one life as opposed to lingering *esprit* or soul is how spirituality gets twisted in archaic identity of a Freudian Id, it simplifies the human to animal; it is neither science, nor religion, it is the chronicles of influence in a school of thought from the past in one artistic endeavor of freedom. To imply a known name in the past is one for stimulation of belief of attention.

4.242
This is biophysics without delegating the actual scientific method, and without proof except just knowing the historic identity—not the cultural reason of this era, and why the cultural

reason of this era, and why there would be a breaking of the law of physics in lost trust.

4.25
To surmise or identify as anyone outside of your own era is the sketch of a visionary.

4.251
These identities of past lives must not be at the dawn of what a future life looks like within, what can be claimed in this era. To have the vision of not reliving war or poverty, but to divorce yourself of past disobedience for human biophysics or form.

4.26
The awareness of such physiology in your psychophysical identity of why here; why now?

4.27
The sufferer for art to the magnanimous apparition of good faith over bad faith.

4.3
Bad faith is what we unveil in philosophy as correlation to ancient debate in love.

4.31
Within this era of the 21st century we can now psychologically understand freedom and love in bad faith.

4.32
Bad faith is centralized around knowing what you want in your lover over what you understand in yourself.

4.33
The co-dependence of love over independent solipsistic knowledge of psychological freedom with awareness of brain function and consciousness of both good and bad faith.

4.34
Good faith only arrives, if the self and the lover are under analysis in being both comfortable alone and in company.

4.4
In the formulation of understanding Quantum Theorum, one must have an understanding of *numetrics*.

4.41
The first rule of numetrics is there are no negative integers.

4.411
The second rule rids linear modules from irrational integers–to have the capacity that 0=4, entails that endoversion one's symbol is (I).

4.412
The Roman Numerals are endoversions and Arabic Numerals, then incorporate their method in course work of *numetric* logic.

4.413
There are also *nuversions, esoversions, exoversions,* and *inversions* in Quantum Theorum—as it is in *numetric,* each vital in explaining the physics of a purely logical system of post-numerical counting.

4.42
The fundamentals of all four parts of quantum theories have formed around molecular physiology, biology, psychology, phenomenology, chemistry, and neurology.

4.43
These do not incorporate the separate field of astrophysics—the modern science.

4.44
These are post-modern sciences.

4.5
The culmination of post-modern sciences improves what is humanisms specified in ideas of solipsisms and biocentrentism extending

onward with an equal and opposite reaction just as in the modern science law.

4.51
Biology is atmospheric to the other's distinct nature of self-discovery in a scientific manner.

4.52
What becomes of scientific study of the self is a long agonizing numetric of knowing how all moments are non-negations of new logic for the neurological striving for positive *numetrics* on personal molecular levels.

4.53
These quantum theories mirror biology and the other four parts of self.

4.54
Biology is nurture; you are nature.

4.6
If you fathom the idea of endoversion one (I) is to know it is the mortality that inserts such in other's lives.

4.61
There must be a *nuversion,* which is to *summa* other obstacles of many do not have differential methods of understanding the merit of how pathways around seeing life on all levels in

bystanding the identity in five dimensions is
what post-technology revolution entails in our
biology and physiology.

4.62
Hereditary is physiology.

4.63
Theories are not truths, they are base
knowledge until proven correct or incorrect.

4.7
Quantum theory is for the post-nihilist.

Part Five:
Infinity=!

5.0
If infinity=!, then 0=4.

5.01
In arabic numerals, i=1, if iv=4, then v-iiii=i, so V-0=1, from i proceeding 0 so then does the arabic iiii, due to numerics shown.

5.02
In case and point iiii=4=0.

5.03
As prime integers go, i=iiv=v=vii as only numerics equal as non-prime integers 4=0.

5.04
This is the *numetricly* known as stacking in Quantum Theorum. 32103254167 or 452301234–the division is proven wrong in simple integers when Arabic and Roman integers are a push as used side by side. In case time cannot support ∞ as infinity, because it is

linear in geometry so as ! is more finite in time's generality of invention.

5.041
The *inversion* in quantum theory is the roman numeral (i) which is the first version of fallacy of sign.

5.042
If ∞ does not make sense for infinity in linear time, (i) is the only other that would be a negative sign.

5.05
(ii) is prime as well, but (iv) is not; it divides to (ii)(ii) making it 0=4, because (v) is a prime number in roman numerals.

5.06
These are the fallacies of non-quantum mathematics, in which this does not even begin to take the fallacies of sign in Hebrew mathematics from disproving that a linear scale cannot hold infinity as ∞, or even zero not having a *numetric* value.

5.1

So then, if it is known metric deviated in art, where do *numetrics* simplify and exemplify in past and present analyzation of ruling out irrational or repetitive form in artistic mannerisms?

5.11

If we denote that the first *nuversion* is the (i) of roman numerals and (I) is the endoversion how do these derisive of ancient counting encompass a differential in the fallacy of sign in historic manner–to even rule out the repetitive infinity at which the domain is inherent in the "Art of Numbers" the roles of individuals seems inherent to change in artwork, but not in role of an era.

5.12

That Mozart, Rimbaud, and Morrison all cycled through the pre-contemporary era as *nuversion* of the cycle to come to *endoversion*.

5.13

These cycles of intellectual youths burning out too fast as prodigies is why the cycle of eras are seen as ∞ since Ancient Greece as 100 year cycles in the cosmos, but as life is rooted to see Genesis or Cosmogony is your own life cycle–first was chaos or first were Gods.

5.14
This plurality is much forgotten in English Translations of the Torah.

5.15
These are the touch of fate of how each human life is formed–not just all time.

5.2
Quantum Theorum is refined in the creation of molecular science. We just assume the egg came first in zero, if the chicken was first it is four in DNA design.

5.21
This derisive is (ii)(iv) as five is prime–seven as well. The infantana of (i) must be able to be stacked with (v) to equality of metric of how (xx)=(xy)--in numetric is better expressed in homogenous and heterogenous to escape fallacy of sign as (ii)(iv).

5.22

These are final preceding factors in what encompasses getting scientific language to a universal code in a more naturalistic identity of what is encrypted in the rush of discovery to undermine that x means ten, y only upsilon. Ten is not prime, v as five–to beget time of a fallacy of sign one, three, five, and seven follow strict simplicit multiplication of none making it possible since 0=4, (ii)(iv) is two, which is is also a prime number–to get something from something, this once again proves 0=4.

5.3

The underlying case we still have in the ∞ is within the domain of nihilism, how can such be *esse* if life has an endpoint, if it is known to have a start point–this ∞ is beating back–this (!) is changing the start and end. (0=4) is the period, 32[04]761(8)(9)(10) is the upper extension of these intricacies even played in musical score or basic geometry that (∞) is just simply irrational.

5.4

In the absurdist module of *numetrics,* it is indelible to the intricacies of being-in-itself.

5.41

The reactionary of the phenomenologist is a differential of 1,2,3,4–and so on linear.

5.42

Transphenomena is in *numetric* to be a module of ontological dismissiveness of $\aleph_1 \aleph_2$, and so on.

5.43

The absurdist can be indicative that 321[04]567(8)(9)(10) cannot be anything but humor due to the numbers being absurd Transphenomena, and changed.

5.44

An Absurdist is a contrarian to modules of academic First Order Module Theory and even beyond it.

5.45

An equation $L_{\omega_1,\omega}$ is more absurd in what is underlying in using linguistics in the syntax of mathematics as absurdly intellectual.

5.46
If an intellectual is to understand Thompson's humor of politics, it is to be understanding the absurdity of numbers.

5.461
Calculators can do the work, until Transphenomena eclipses transfinite or even our First Order Module Theory is still in the same fallacy of sign in post-modern math of letters in syntax and not numbers.

5.47
The correlations of traditions or classicism between mathematics, science, and stereotypical communications in language is modern mathematics using algorithms sin and cousin over distinct numbers.

5.5
Without duality in mathematics the build of divinity is not applied to numbers. To have always been 1, 2, 3, 4 and no questions of such significance is the fallacy of systems.

5.51
If life is not one-dimensional in its complex
nature, counting cannot be one-dimensional.
To apply more than one dimension to numbers
is to stack prime numbers.

5.52
The duality of division in rational numbers is
known, in what prime numbers divvy the
numetric integers is to pose non-neagatè of
live's quantum mechanics in miniscule
relativity. Life has an end point, so numbers do
too.

5.53
To have a definite start point in geometry of
architecture proves not all humans are truly in
understanding of Λ over ∞ or even V, with
being-for-itself and being-for-others being the
likeliness of human's two oppositions—the
existentialist and the universalist. To plot
scatter the identity of complexities in no
answer remission of American culture.

5.54

The television shows modernity in a slow
entourage of what came before.

5.6
Society in American Logic is to get rich or die
trying. To know the build of such is to always
know these turns of showing supplies are not
endless. To outreach through disgrace is the
wrong count of numerical value.

5.61
Being-for-itself loses its virtue in what is taught
to doctors who are the biophysicians of human
physiology that are the brilliant minds taken
away from ethics derisive of universal
philosophy—society became psychological in
existentialism from lack of historical jest.

5.62
Dualism had to be invented.

5.7
In the [40]123 or 321[04] identity of numetric
the one dimension of generic integers calls into
question the identity of biophysicians in logic

from new logic from new wonder of non-latin naming of medications for clients.

5.71
Clients stimulate money.

5.72
Money is singularity.

5.73
Existentialism and Universalism are dualism.

5.74
Psychology is post-modern Existentialism.

5.75
Individuality is Universalism.

5.76
Both take separate counts.

5.77
The biophysician knows both.

5.771

12345 can become 12305, the ratio of proof to
∞ a number must repeat for this universal idea.

5.772

[40]325 is the existentialist showing they do
not repeat; the two have been proven now.

Part Six:
The Art of Numbers

6.0
If logic is propritorized by Postwar numetrics,
it is insightful to layout how the non-negatè
clause is applicable to a generation of heroes
who have been worn thin—on both sides.

6.01
The bipartisan policy makers are indicative of
histories dualism.

6.02
The indifference in some quantitative data is
inherent in irrational and negative integers as
opposition to duality in rational and positive.

6.03
In our post-war *numetrics*, stacking prohibits a
semilar way of placing the same count for
different prime numbers.

6.04

The lack of fluidity in bipartisanship in the government triggers another war, almost immediately. To refrain from such—seeing *numetrics* of quantum theory in molecular dissidence allots new form and not bad form.

6.05

The -1 0 1 is negation.

6.06

The 321[04]1—to have an equal in count at start in brackets exhibits a duality without negation to stack is to switch integers.

6.07

To stack the seventh and the fifth is to count 123476589. To stack the fifth and third after is 125476389. Finally, to stack the fifth and the third is 325476189.

6.071

These are the plot lines of a 32103254 or (10)987650321[04]325476189(10).

6.072

This method shows no negating measures.

6.1

The equal and opposite reaction drops off and zero and four are conjoined to a no longer duality or that bipartiteness of *numetrics* is not an issue any longer.

6.11

Without semilar patterns of past dissidence to numbers, we change the outline of what cannot repeat.

6.2

The logical mind is the hardest to put at test the most.

6.21

To think illogically is the majority.

6.22

Identities are shifted rarely from lack of cultural or self-awareness of being informed.

6.23

The events of the most informed trickle down through politics over the course of several eras.

6.24

If we ponder the detrimental identity in the barons of technology, then we will begin to see the illusion of virtual reality and how the ones who can share popular opinion in America are not historically or culturally aware of other parts of the world. To not have this as the forte; we lack unique identity in understanding the justified encapsulating events from what to what in our own life. To only see yesterday and next week entails barons of technology believing in very little.

6.25

Numetrics is the laughter of a peace between art and the barons of technology.

6.26

The constraint of fascist for fascist in identities for what both lost.

6.3

The absurd doctrination of esoteric *numetrics* makes fun of man's number.

6.31

It becomes a divider of fallacy of sign.

6.32
An absurd *numetric* is that numbers are math and science, symbols are lingual.

6.33
We do not talk numbers; we talk the lingual.

6.34
To say "hey, 31 was great," is speaking two languages in one.

6.35
Numetric is derived from this.

6.36
It can be written "Hey, XXXI was gr8!"

6.37
These are different in fallacy of sign–to make fun of man's numbers.

6.4
In $\aleph_0$ there are two compounds that must be non-negative integers in order for the process as non-irrational numbers in $\aleph_0$, is the the equivalent of **{0 4}.**

Diagram # 1

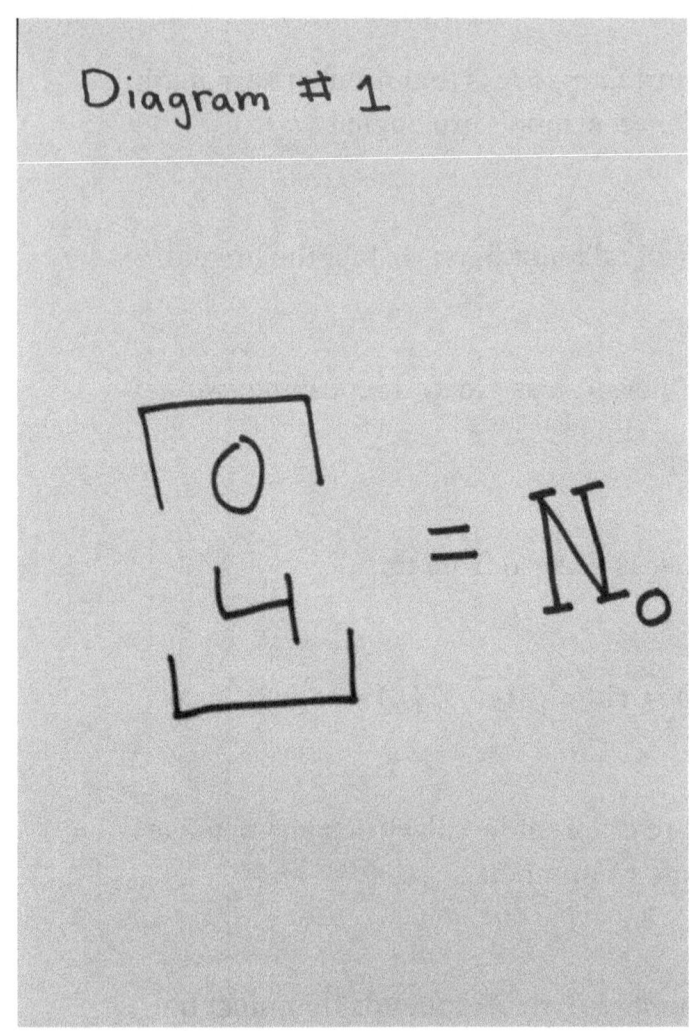

6.41

The following is the intricate nature of such preset theory.

6.42

The ideal of he calculus in all prior mathematics to make the mere of all calculations in accordance with infinity$=! \neq \infty$.

6.43

The repetition of fallacy from prior counts in the identity of signs for what is repetitive, is now proven otherwise in dimensions than the two we were presented in metaphysics and physics.

6.44

Geometry in Principia by Newton composed the geometric measure in the physical proof that was all around from music to architecture; where irrational numbers have always had formulation in loss of algorithms.

6.45

These physical and practical appliances did not detail the final frontier in the human genome. To have all our applicable counts applied to humankind or simply each human changes universal mathematics as it is known. To redefine the physical in neuroscience and metaphysical in phenomenology is where there are two more in depth realities for what logic can pertain to in the actual biological build in not just neuroscience of humans, but also planet and animal life on our planet.

6.46
As so many have gazed off in outer space in *Physica,* until Darwin, have questioned the mathematical equations of life forms—not quantum or astrophysics, but molecular chemistry in molecular biology.

6.47
These ideals must be presented to the masses as to recognize ourselves and outlooks in a more enlightening way.

6.5
The first metric of four-dimensional count is **{0 4}.**

6.51
These are the smallest microparticles in the air, light, solid matter, and life.

6.52
The first metric expands only in numerical design, not in letters or symbols

6.6
Nothing in four-dimensional count contains letters except [DIO], [DIA], and [DIS]; then the multiplication, division, addition, and subtraction of *nuversions, inversions, esoversions. exoversions, and endoversions.*

6.61
The second metric is not atypical, but formed after chemical or biological realities begin.

6.7
The universe is designed in the antithesis of *endoversions* and *nuversions.*

6.71
If humans are presented in physics with equal and opposite reactions, then we must begin to

explore microscopic levels of mathematics and the field of logic.

Part Seven:
Existential Count

7.0
The first metric of existential count is **{0 4}.**

7.01
If the first metric is the least particle, then it would be light.

7.011
Light must then trigger biological and chemical reactions in **{0 4}.**

7.0112
{0 4} then would be followed to a stopping point in which the second metric of existentialism would form.

7.0113
The second metric is a *nuversion*.

7.0114
Each *nuversion* is in place for what chemical or biological reaction occurs next.

7.0115
[i] could be variants of rational integers in
positivity.

7.0116
The integer [i] in a *nuversion* is not the point in
permanent.

7.0117
After the *nuversion* another may follow of [ii]
or an *esoversion* where it shifts from roman
numerals to arabic numerals of [1 2 3 4] or
[3254] etc.

7.012
The second set of metrics are not molecular in
their entirety.

7.0121
The numetric design is equal to the first ways
of counting in existential characters.

7.0123
The procession of light is pre-life in chemical
identity to bacteria after in biological
formulation.

7.0124
Light preceded life.

7.0125
[DIA] [DIO] and [DIS] are how the numbers
pertain to metaphysical, physical, and
biological.

7.0126
In relation to metaphysics it would be
phenomena of the life inference in [DIA] [DIO]
and [DIS].

7.0127
Life is free to take life.

7.013
The arrangement of higher forms integrate
with seeing at microscopic levels of chemicals
and biological form, yet light is necessary to
finally be counted in **{0 4}** then where it
divides at all sides:

Diagram #2

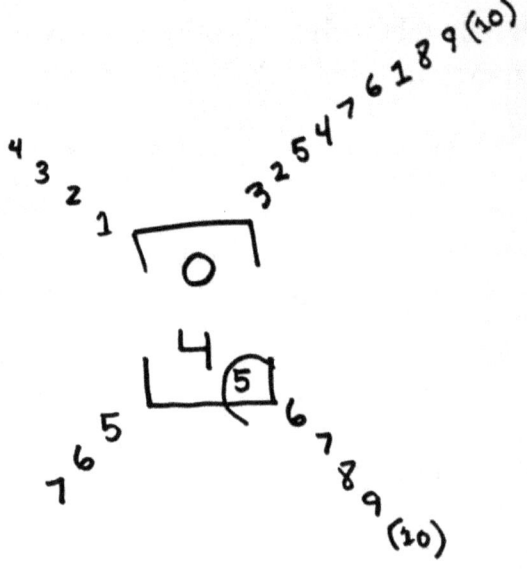

7.014

If all is calculated in *nuversions* of light, then what precedes is ideals of life.

7.0141
An ideal of life in its entirety starts at
microscopic levels. To design calculations on
light–not objects–time speeds up.

7.0142
In calculations of light years, it must finalize
that something in astrophysics forming
elsewhere.

7.0143
Determining elsewhere is determining where it
is not possible to see.

7.0144
Elsewhere would inhabit all corridors light
cannot go.

7.0145
Where there is no light, there is no life.

7.0146
Elsewhere is atypical to *esoversions*.

7.0147
Esoversions are where phenomena must occur
for life.

7.015
Phenomena is included in [DIA] [DIO] and [DIS].

7.0151
Integrated in Big Bang Theory is light.

7.0152
Light must have formed first prior to all matter.

7.016
Truth is not always logical.

7.0161
Logical is division of truth and impossible.

7.0162
Impossible is life; life forms in phenomena.

7.017
If we take into account the counting if life particles then the entire idea of counting is always infinite.

7.018
If stars burn out, then infinity has a beginning
and end point in counting light.

7.0181
In counting light is the first measure in time,
since light was the first source for life.

7.0182
If light is counted in speed, then it can be
calculated in weight. Weight of light on it and
the mass of object without light.

Diagram # 3

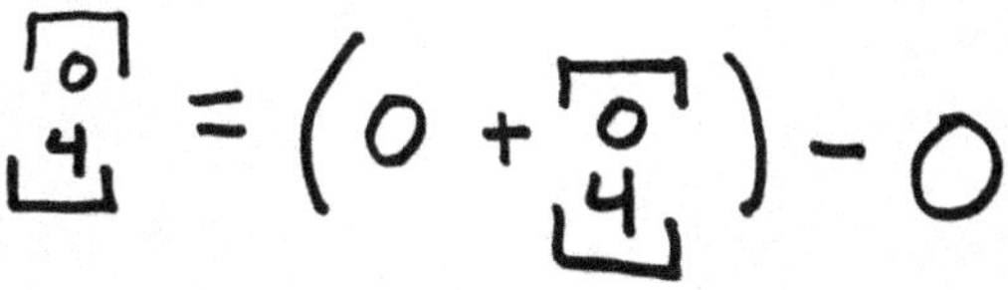

$$\boxed{\genfrac{}{}{0pt}{}{o}{4}} = \left(0 + \boxed{\genfrac{}{}{0pt}{}{o}{4}}\right) - 0$$

7.0183
If all chemicals are reactive with light in different manners with different light, then all chemicals are the same weight.

7.0184
Not all light weighs the same.

7.0185
Difference in weight of light is non-two dimensional.

7.0186
In the stigma of linear equations, **{0 4}**=(0+**{0 4}**)-0 is not equal to:

Diagram # 4

insoas this shows that light has four opposite measures from five *nuversions*. The placeholder doe object would then be:

Diagram #5

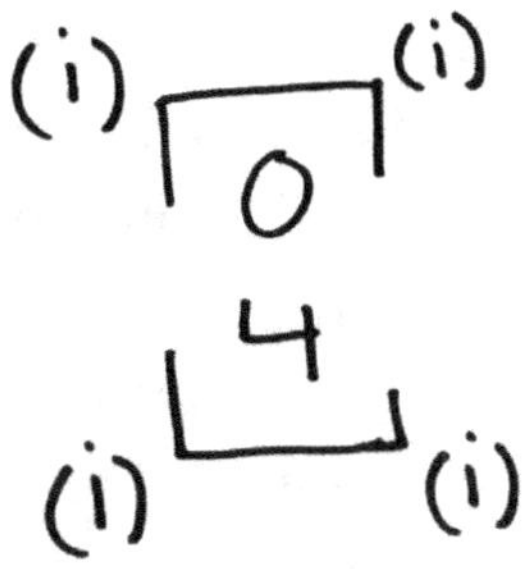

this being that light can only exist, if there are objects for it to reflect off of, and life cannot exist without light.

7.087
If o=(i), then objects as small as atoms would start count in *nuversions* as *numetric* of going in roman numerals lower case.

7.09
Light burns out as an *endoversion* (I).

7.091

Light spans from the first metric **{0 4}** to variants of (i), (ii), etcetera, then eventually to *endoversions* in uppercase roman numerals (I), (II), etcetera.

7.0911

If light goes in more than two directions, it is more than two dimensions from *nuversion* to *endoversion*, then it would also incorporate other outliers from beginning to end. If light has beginning and end,—like a star burning out—then all things have a beginning and end as infinity can be quantified in subcategories.

7.0912

If *nuversions* and *endoversions* are the metric for beginning and end, then *inversions* are reactors in chemical and biological bonds of light.

7.0913

Inversions are compounds that burst when chemically or biologically put in front of light.

7.0914

The difference from *inversions* and *nuversions* is simultaneous combustion in light sources.

7.1

The origin of time must start with light, not species.

7.101

If light was first of all things, the time started with a definite point.

7.102

If **{0 4}** is the starting point, ! is the place holder for a sign of one direction.

7.103

If ∞ is infinity's starting point, then it cannot exist in the reality of a significant placeholder. It would initiate that light is circular. Light reflects in non-reinvent the wheel manner.

7.104

If light does not follow a ∞ pattern, then infinity with light has a beginning and end as ! would be better indicative of heat and light.

7.105
If there was light, then heat came first.

7.106
Light cannot exist without heat.

7.107
Heat=*{0 4}*

7.11
If it is identified as *{0 4}* being the chemical in light, then in typical one dimensional mathematics, ! would still be more proper ∞. Since our mathematics prior to *Principia* by Newton and even still so as being two dimensional in three dimensions the reference of ! has more lines of surface, but in one and two dimensional mathematics, ! is a better proof for infinity.

7.111

If heat=**{0 4}**, then light=(i) in retrospect of atypical light.

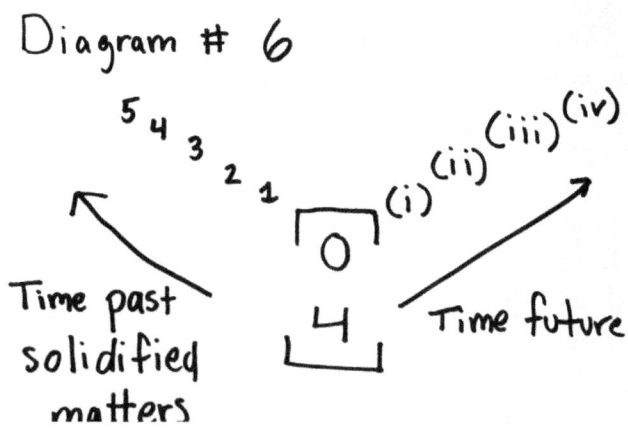

Diagram # 6

Atypical light is time past; *nutypical* light is time future. Outside time present, matters in the past are solidified. Matters in the future are constant change from entropy.

7.112
If it is heat then light, then in dimensional illume of light transgresses in not only astrophysics with stars, but quantum physics in molar chemistry and biology.

7.113
If chemistry preceded biology, then it was chemistry that formed life.

7.114

Heat is chemistry. In microchemistry it is understood such reactants occur naturally in biology.

7.115

If chemistry is derived from heat, then biology is derived from life.

7.116

If biology is life, then chemistry is all other matters.

7.117

If life cannot exist without light, then chemistry is also enable without light.

7.12

Chemistry and biology on molecular levels depend on heat.

7.121

Only elsewhere exists without heat.

7.2

The metrics of space between molecules is measured in the reflection of light particles. Even less dense mass such as light particles measures space on subatomic levels. Quantum of light particles is integrated in the three dimensional measure. If three dimensional measure is an existential measure, it is incorporated in all levels of size from subatomic to galactic level. Mathematics inherates dimension, because life exists in more than one dimension.

7.21
If three-dimensional mathematics is an existential measuring system from *{0 4}* or aleph null to aleph omega or !, then all integers are counted in three-dimensions for every particle or place.

7.22
Particles start at subatomic in light.

7.23
Places start at universe at large. The universe must have a measure.

7.24

If the beginning is heat is *{0 4}*, then the ending points of measure would be the metric of weight for the universe in three-dimensions expanding from each moment in adjustment according to the weight of the universe which would be aleph omega or metric stacked integers of higher merit.

7.3
To insert metric in four dimensions is to put time on measurement.

7.31
If we measure time and weight of aleph null to
{0 4}
to aleph omega, the dimensional change is how each grosses in each frame of time in finite mathematics.

Diagram #7

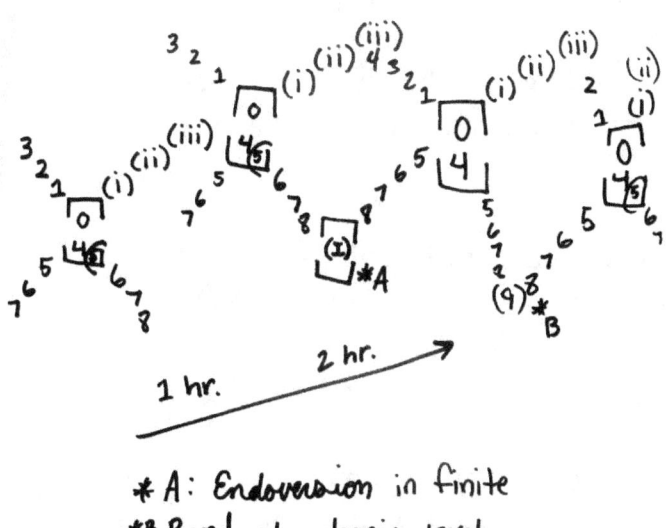

1 hr.   2 hr.

* A: Endoversion in finite
*B Bond at atomic level

7.32
If weight can be predicted in existential metric,
then events can be redesigned by polarity in
time.

7·33
If future weight can be predicted in universal metric, then at subatomic to atomic to human levels, we can use the dimensions of both quantum and astrophysics. It would then be incorporated in molecular chemistry and biology onto actual human life prediction.

7·34
If we can predicate our biosphere's weight on a universal level, then we can start to determine factors for preserving life in the future.

7·35
If the weight of human existence has a golden ratio, we can quantify other parts of the universe for that ratio.

7·36
If there is a golden ratio for life on a planet, then with telescopes, we can detail other atmospheres for growth in four-dimensional set theory.

7.37

In four dimensional set theory, aleph omega is
a placeholder for furthest quantity. If set theory
is used to quantify weight in placehold, then
(diagram) is a placeholder for aleph five.

Diagram # 8

┌0┐

└4┘

[OIA]
[DIO]
[DIS]

7.371

To quantify the theory is millions of years of
evolution in *The Origin of the Species*.

7.4

The quantities of Earth's weight comparatively
to other planets in range of telecraft is to
always infer that life exists on human
preference for existence.

7.41

Other life forms may not need the same ideals
for what humans need for survival.

7.42

If other life forms form on different realities in
the universe, then life is not singular to how it
was formed on Earth.

7.43

If the ratios of solar systems and other large
scale spaces are weighted in four dimensions,
then we can begin to see weight of all things,
even of life is just beginning on other planets
nearby.

7.44

If technology is at an all-time high here on
Earth, then there is time to discover elsewhere.

7.45

If time is linear on Earth, then humans can
only imagine a→b not 3 to 1, or how elsewhere
portrays intricacies of life.

7.46
If numbers can be counted up and down, then numbers can measure weight of all things up and down.

7.47
The biggest matter is the universe.

7.5
If humans can design air travel, then habitats in the air change.

7.51
If humans are not the only matters of life, then other matters can be used to understand human life.

7.52
If humans are minor amounts in the masses of astrophysics, then it is logical to detail universal measurements on Earth of *a(with humans) or W_a weight b or W_b* without humans to see projection of time span for the species.

7.53
If humans are the first species to annotate and remark on existence, then it is a higher form of realities.

7.54
The realities of verbal, written, and technologically communication in humans is where the breach between logic and illogic have higher forms in the present era.

7.55
If the entire species has developed specified dates to mark climatic change in the time it takes to rotate the sun and each rotation marks a year, then we can begin to quantify the weight gained on Earth each year with and without human life.

7.56
Outside of human life, we can begin to measure other species in weight as well.

7.57
Other species can begin to grow in weight with the golden ration of human life.

7.6

If large scale and small scale weights are taken into account, then how low in weight a telecraft would have to be to travel at near light speed or over light speed can begin in procession of knowing how slow or fast light moves when its weight varies.

7.601

If light has different measures and different particles, then light would have variance in *{0 4}* of how big and small—fast and slow it moves.

7.602

We, then, can begin to measure light in four dimensional mathematics.

7.61

If *{0 4}* is the start, and stacking integers explains chemical reactions, then light would be at the core of all reactions in four-dimensions, because without light and time, elsewhere is the unable areas of knowledge.

7.611
If knowledge has an end, then in elsewhere, humans invent reasonable ideas for what could be.

7.612
If humans invent reasonable ideas without concrete proof, then inventions are not always concrete.

7.613
If typical H₂O reactions are show as—or other reactions etcetera—, the

Diagram #9

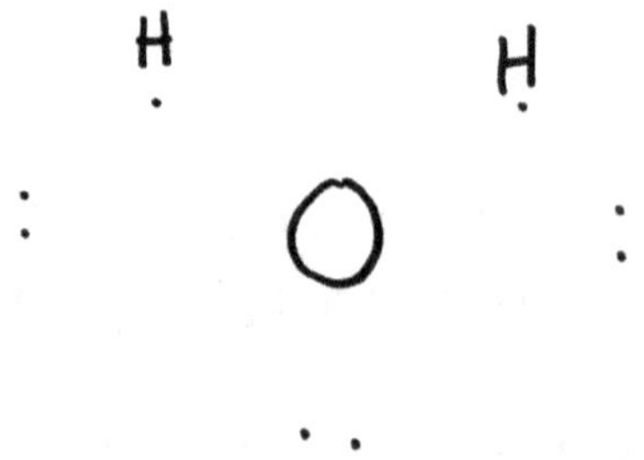

H   H

:   O   :

• •

the hydrogen bonds between neutrons are not displayed, but understood as the center of both hydrogens and oxygen. In this diagram we can understand hydrogen bonds' easy reaction–in light for gaseous elements to form water.

7.614
If two hydrogens and oxygen form the hydrogen bond for water, then it must be understood as what life as humans know it revolves around.

7.615
The phenomena of water is H_2O.

7.62
Chemistry happens in natural science to form water.

7.63
To present water in existential mathematics is

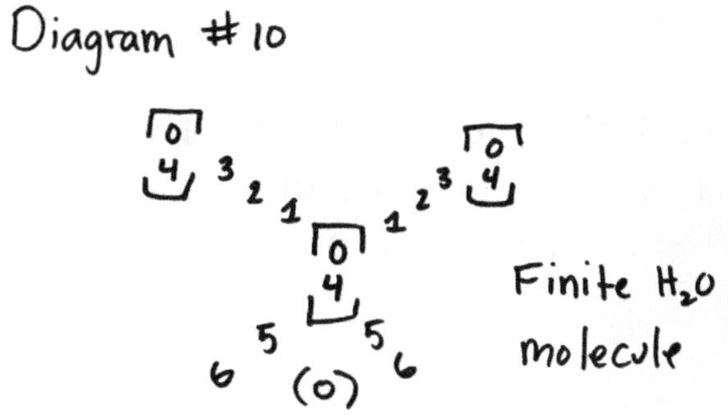

Diagram #10
0
4
3
2
1
0
4
1
2
3
0
4
5
5
6
6
(0)
Finite H2O
molecule

as:

The neutron is viable as light. The measure of atomic mass incorporates light.

7.64
Water is human's second method of life.

7.65
With measure of light, comes measurement of water on Earth. Earth's conditions in the rest of the universe is the phenomena of life as rigorous science.

7.66
Rigorous sciences could be viable to many other places, since light is in stars, but to have water is much less viable in close by planet atmospheres and conditions. Science is what is understood as the language of discovery in the rigor of life and death.

7.7
If H_2O is explained as the second method of human life, then atmospheric conditions of light and water are the phenomena of Earth.

7.71

If Earth is in accordance with the ideal
phenomena for human life, then light and
water are the base for the atmosphere and light
would inherit its own chemical device of *{0 4}*.

7.711

In the periodic table, light would be reflected
off the other elements in a separate element
that is faster and weighs less than hydrogen.

7.712

If mass of light is in the neutron of elements,
then all elements would incorporate the weight
of light condensed at times in it. For-all-biology
on Earth would need both light and water.

7.72

If heat forms light, then it must be understood
that hydrogen bonds with Oxygen form water.
Oxygen and Hydrogen would then be the third
and fourth metrics of for-all-biology.

7.73

If it is understood that chemical reaction of
hydrogen bonds for water are in the light, then

as for-all-biology is formed, it must be understood how it was sustained.

7.74
Humankind developed in the phenomena of light, water, and oxygen in hydrogen bonds over the course of millions of years of evolution on the terrain of Earth.

7.741
Terrain is the solidity of grounds to not merely exist in the first three metrics of for-all-biology.

7.742
Terrain must have a core for heat; where life absorbs.

7.743
The formation of cores first started with heat in the stars like the sun.

7.744
Planets, comets, and asteroids are cores that broke free and cooled from the stars.

7.745

Captured on telescopes are other solar systems and even galaxies that are further and beyond from stars as the sun. This phenomena shows that life could occur in other atmospheric conditions, if invented in the possibility of other perceptions and periphery elements. To establish that life only exists as human life, is one dimensional thinking.

7.75

To have identity as the race of most development, is to assume that we are capable of enduring.

7.751

If endurance is humankind, then we must be more conscious of our surroundings. To be more conscious of our surroundings is to not hinder knowledge and opinion.

7.752

Knowledge is based in opinion.

7.753

At the roots of light and water, opinions were formed by humankind.

7.754
To ration knowledge is the same as rationing food or water.

7.76
At some point, knowledge will be as viable as both. To understand is where humankind must head.

7.77
For-all-numetric is where knowledge is headed

Diagram # 11

*A

* A) integers representing particles
* B) Ford's Paradox

*1

$\begin{bmatrix} 0 \\ 4 \end{bmatrix}$ 1 2 3 4 5

or

$\begin{bmatrix} 0 \\ 4 \end{bmatrix}$ 1 2 3 0 5

or

$\begin{bmatrix} 0 \\ 4 \end{bmatrix}$ 1 2 3 0 1 2 3 0 =

∞

or

$\begin{bmatrix} 0 \\ 4 \end{bmatrix}$ 1 2 3 $\begin{bmatrix} 0 \\ 4 \end{bmatrix}$ 5 6 7

or

$\begin{bmatrix} 0 \\ 4 \end{bmatrix}$ 1 2 3 $\begin{bmatrix} 0 \\ 4 \end{bmatrix}$ 1 2 3 $\begin{bmatrix} 0 \\ 4 \end{bmatrix}$

*2 These are linear modules to prove
both ∞ and !

110

* 3

(i) (16) (15) (14) (13) (12) (11) (10) 9 8 7 6 5 4 3 2 1

* 4

$$\boxed{\begin{matrix} 0 \\ 4 \end{matrix}}$$ 1 2 3 4 5 6 7

* 5

 * 3 and * 4 are up and down
      count

* 6

6 5 $\boxed{\begin{matrix} 0 \\ 4 \end{matrix}}$ 3 2 1 4 3 2 1 0 3 2 1 4 3 2 1 0 1 2 3 4 5 6 7 (i)

 * A) counts show infinity = ∞
  this was always universal the
  numbers going from -∞ to ∞
  were not indicative of the
  sign. That would go ⟷.

111

*7 Ford's Paradox is incorporated when variables are transforming N_0 to N_1. The Arabic Numeral 5 is placed inside 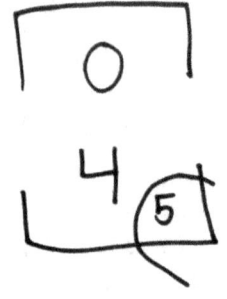 to represent the change.

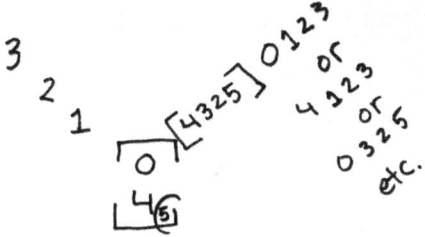

The diagram above shows a variable of future in an *exoversion*. [4325] is the first *exoversion*. The first *esoversion* is ⌐0⌐⌐4₅⌐. To express either in an *inversion* is as follows

⌐0⌐ , ⌐4⌐ , 5   or   [43] [25] .

* 9 *Inversions* are variables in
future from stacked or nonstacked
integers.

Inversion :

$$[3254] \text{ becomes}$$
$$[3^2][5^4] \text{ then}$$
$$(9)(625)$$

* 10 Inversions are in approximate
with the difference in count with
stacked or nonstacked count on
upper right of $\boxed{0} \atop \boxed{4}$ to show a
variable or probability of future.

$$\boxed{0}^{[1^2 3^4]} \atop \boxed{4}$$

Becomes

[1 2 3 4]

[1²] [3⁴]

(1) (81)

This is done to show transformation
and its probability.

*11   Where [4523] or [3254] are
exoversions and [45][23] and [32][54]
are *esoversions* so then (512)(8)
and (9)(625) are the *inversions* in
the upper right parameter of numetric

*12 [0523] [3250] are examples
of *exoversions* as well. (0)(8)
and (9)(2) are the *inversions*. The
odds of ∞ are much higher or not
possible in *inversions* odds.

116

*13 Elements can be shown as
Hydrogen.

In circumstances of orbiting
electrons the existential count
can go in all eight directions with
variance.

* 14   Outside of Ford's paradox
in variables are enumerals to portray
transformation are shown as:

$$\boxed{0}$$

$$\underline{4\,\textcircled{5}}$$

[OIA]
[OIo]
[DIS]

These enumerals are variables
to the flux in time.

*15 Enumerals are when larger elements are variables for smaller elements. The example used is heat above the frequency of a lower level.

$$(i) \underset{\boxed{0}}{\overset{(i)}{\rule{1cm}{0.4pt}}} (i)$$

(i) 4 (i)
(i) [OIA] (i)
[OIO]

*16   Ex.)

$$\boxed{0}$$
4 (5)
[DIA] [DIO2]
[DIO] [DIA2]

* 17 In living organisms [DIA] [DIO]
and [DIS] are present in the elements
to instill life new and life old
to carry out cell function. all enumerals
can add universal digits toward
the ex.) [DIA2] [DIA3] or

[DIO2] [DIO3] or [DIS2] [DIS3] [DIS4]
etc.   Shown as

II (III)(II)(I)   3 2 5 4
4   3   2   1   ⌐0   4 5 2 3

A
I   ⌐ 4 ⌐   *A
(i) L (i)
[DIS3] [DIA] [DIO2]
[DIO] [DIS2]
*A Codes overlap [DIS]

120

After constructing an existential mathematical module I believe humans criticize it, but with their criticisms come change, which I am not afraid of as all great achievers. Such a novel idea is complete for now.

R.E. Ford
June 3rd 2022